I0066955

Byron Briggs Brackett

The Effects of Tension and Quality of the Metal

Upon the changes in length produced in iron wires by magnetization

Byron Briggs Brackett

The Effects of Tension and Quality of the Metal
Upon the changes in length produced in iron wires by magnetization

ISBN/EAN: 9783337125363

Printed in Europe, USA, Canada, Australia, Japan

Cover: Foto ©berggeist007 / pixelio.de

More available books at **www.hansebooks.com**

THE EFFECTS OF TENSION AND QUALITY OF THE METAL

UPON THE CHANGES IN LENGTH PRODUCED IN IRON WIRES

BY MAGNETIZATION.

by

Byron Briggs Brackett.

--o--

Submitted to the Board of University Studies of
the Johns Hopkins University, as a Thesis for the
Degree of Doctor of Philosophy.

May, 1897.

--oo--

THE EFFECTS OF TENSION AND QUALITY OF THE METAL
UPON THE CHANGES IN LENGTH PRODUCED IN IRON WIRES
BY MAGNETIZATION.

This Investigation was suggested by Professor Rowland
and has been conducted throughout under his direction.

-oo-

HISTORICAL.

Fifty years ago a machinist of Manchester imagined that he could see the volume of a mass of iron increase when it was magnetized and decrease when the magnetizing force was removed. Hoping to be able to use this principle in the construction of an electro-magnetic engine, he appealed to Dr. Joule[1] to investigate the phenomenon and determine the amount of the change. By immersing the iron to be magnetized in a closed vessel filled with water in which stood a fine capillary tube Joule could not detect any change of volume, though it has since been shown that had he used[2] either stronger or weaker fields he probably would have done so. But by a system of compound levers of great multiplying power he proved that an iron bar did change its length when magnetized longitudinally. He observed an increase in length of 1-200 000. As a result of his investigations he proposed the following laws:-

1. When soft iron rods are magnetized their length increases and the elongation is approximately proportional to the square or the magnetizing force.

2. Tension applied to the rod diminishes the elongating

----

(1) Joule. Phil. Mag. (3), vol.30, pp. 76,225.
(2) Bidwell. Proc. Roy. Soc., vol. 56, p.94.

effect.

3. The elongation is greater for the same intensity of
magnetization in proportion to the softness of the metal.

That the two first laws are correct for the fields
that Joule used cannot be doubted; but as to the third law
there is much uncertainty. The investigations of Shelford
(1)
Bidwell indicate that not only hardening but also annealing,
an iron rod diminish the elongating effect, and at the best
the relation between the softness of the iron and its
change of length is to-day very much confused.

It was nearly twenty-five years after Joule's investigations before the question was taken up again experiment-
(2)   (3)
ally by Barrett and nearly at the same time by Mayer. Barrett employed the tilting mirror, which is described under
the apparatus used in this investigation, a device suggested
to Barrett by Professor Rowland. He experimented not only
(4)
upon iron but also upon nickel and cobalt. He observed an
elongation of 1-260,000 for iron and 1-425,000 for cobalt;
and a contraction of 1-130,000 for nickel.

----

(1) Bidwell, Proc. Roy.Soc. vol. 55, p. 228.
(2) Barrett, Phil. Mag. 1874, vol,47, p.51.
(3) Mayer, Phil. Mag. 1873, vol.45, p.350.
       "   "   "   "   vol.46,p.177.
(4) Barrett, Nature, 1882, vol.26, p.585.

Mayer found an elongation of 1-277000 for iron. Some of his observations upon the action of hard steel seemed to be at variance with Joule's results. But Bidwell has since shown that this apparent difference was due solely to their different methods of experimenting.

(1)

In 1885 Bidwell reported the first of a very extensive series of experiments upon the distortions caused by magnetization in iron, nickel, and cobalt. He carried his investigations up to fields many times stronger than those used by the earlier investigators. He has worked upon the effects of tension ,tempering and annealing. He has experimented with both rods and rings. He found that,at least with his apparatus, rods did not continue to elongate but reached a minimum length, then gradually shortened until they had less than their initial length, apparently approaching a limiting value asymptotically. Investigations upon

(2)   (3)   (4)

this subject have also been made by Berget, Bock, Jones,

(5   (6)   (7)

Knott, Lochner, Nagaoka;   and two years ago investigations

----

(1) Bidwell. Proc.Roy.Soc., 1885, vol.38, p.265; 1886,vol. 40, pp. 109, 257; 1888, vol.43, p.407; 1890, vol.47, p. 469; 1892, vol.51, p.495;   1894, vol.55, p.228; 1894, vol.56, p.94;   Phil. Trans. Roy. Soc., 1888, vol. 179 (A), p. 205.
(2) Berget. Comp. Rend. tom. 65. p.722.
(3) Bock. Wied. Ann., 1895, vol.54, p.442.
(4) Jones, Phil. Mag., 1895, vol.39, p.254.
(5) Knott, Phil.Mag., 1894, vol.37, p.141; Proc.Roy.Soc. Edinb'g., vol.18, p.315; vol.20, p.290; vol.20, p.334.
(6) Lochner, Phil.Mag. 1893, vol.36, p.504.
(7) Nagaoka, Wied. Ann., 53, pp. 481, 487, 487.

in this line were begun in this laboratory by Dr. L. T.
(1)
More.

At Professor Rowland's suggestion, Dr. More determined the intensity of magnetization in his wires for each
change of length observed, and also sought to take into consideration the secondary actions that might affect the
length of the wire.

Professor Rowland defends his position on the subject as
follows:-

The change in length may be partly due to other
causes than the magnetization of the metal. Among these one
can put the stresses due to magnetization. It is not at all
evident that those stresses can be exactly identified with
the Maxwellian stresses. If we think of the long wire that
More used as composed of a bundle of small elementary magnets tied together at points well inside the poles, the magnets would seem to have no tendency, in their central parts,
to separate from each other; and in that case there would
be no pressure at right angles to the lines of induction,
unless it can be shown to result from a squeezing outwards
of some kind of matter, caused by a longitudinal compres-
sion.

At the same time, the compressive force which, it is known, will tend to close up a very thin air-gap in a divided magnet must also exist in any magnet, for according to all our ideas of matter, there is no real difference in the case where the air-gap exists and where it does not; because we still must consider the gaps between the molecules. If we now think of the long elementary magnets as composed of short elementary pieces with their ends so near together that the effects of their poles are neutralized in all action upon external bodies and yet allow a space between their ends for a compressible medium to entirely surround them, then this longitudinal pressure would cause both longitudinal shortening and a pressure outwards perpendicular to the induction. It would seem that such a case might represent both the strain in the medium and the strain in the ether.

The value of this compressive force is probably $\frac{B^2}{8\pi}$. It may possibly be $\frac{(B - H)^2}{8\pi}$, but in most cases the two are so nearly equal in value that it would not seriously alter the results to take either force. Taking this as equivalent in its action to a simple mechanical pressure and considering, with it, the ordinary elasticity of the wire, the shortening due to this cause can be computed for each observation and corrections may be made accordingly. Or this force might be considered to neutralize a portion of the tension on the wire, equal to it in value.

Again, if the magnetized rod or wire extends beyond the magnetizing solenoid then the poles of the magnet will tend to draw into the solenoid, while if the magnetized rod were considerably shorter than the solenoid it seems evident that the poles would have some tendency to move out towards the ends of the solenoid, and thus diminish the compressive force.

Thus, in every case like that of Dr. More's the effects of magnetization must be analyzed into direct and secondary actions in order to get any true idea of the real phenomena.

For reasons like these, Professor Rowland advised Dr. More to correct his observed readings of elongation for a shortening caused by the force $\dfrac{B^2}{8\pi}$; and those corrected readings were plotted to a basis of induction in the iron instead of being given on a basis of the magnetizing force, as all previous curves in this subject had been given, except those of Nagaoka.

Last year Dr. E. F. Gallaudet[1] made an investigation on this subject in this laboratory. He used the apparatus employed by Dr. More, and his curves were in all respects

----

(1) Gallaudet, J.H.U., Thesis, June, 1896.

except one similar to those obtained by Dr. More. Bidwell
and others. He did not correct them for the $\dfrac{B^2}{8\pi}$ contraction.

An initial contraction was observed by him that no one has announced before, and he also gave observations that seemed to indicate a very great change in the values of Young's modulus as the magnetization increased. That there could be no real change in the elasticity so large as these, can be shown from Dr. Gallaudet's own tables; for such changes in values of Young's modulus under the larger tensions used would have caused changes of length in the wire many times greater than the changes observed. Moreover, Bock found as a result of his experiments that the changes [1] in elasticity due to magnetization could not exceed one-half per cent., and Miss Noyes, in a series of experiments upon [2] the changes of Young's modulus with temperature, did not find any evidence of a change of the elasticity from magnetization.

OBJECT OF THIS INVESTIGATION.

Dr. More studied but one quality of iron. Dr. Gallaudet experimented with only one iron wire and one nickel wire. It was hoped that by carefully observing the behavior of quite a number of different wires under varied physical conditions, data might be obtained that would explain the relation of the elasticity of the metal to its change in length, under magnetization. It was also hoped that incidentally the question of an initial contraction preceding elongation, as observed by Dr. Gallaudet, might be settled and that likewise some additional information on the relation of the temper of the metal to its change of length might be obtained.

APPARATUS.

I employed the apparatus used by Dr. More and by Dr.
Gallaudet. It is very fully described by Dr. More who (1)
gives all the dimensions essential to its construction. The
feature that is essentially characteristic of this apparatus, in addition to Professor Rowland's tilting mirror that
both Barrett and Bidwell used, is the cylindrical jacket
which was suggested by Dr. Ames. The figure here given
will show its use. About the part of the wire where the
changes in length are to be observed is placed a brass tube
having a free internal diameter of over one centimetre. At
the bottom end <u>b</u> the tube is clamped firmly to the wire
while the upper end has a loosely fitting cork merely to
keep the axis of the cylinder and the wire coincident. To
the top of the cylinder is attached a projecting arm that
carries two raised supports <u>a</u> and <u>c</u>. Above this arm is
placed a lever <u>d</u>, resting by a knife-edge on the support
<u>a</u>. This lever also has an inverted knife-edge but a very
short distance back of the one that supports its weight,
and over the inverted knife-edge is placed a hook that is
firmly clamped to the wire at <u>g</u>. Thus any change in the
length of the wire between the points <u>b</u> and <u>g</u> will cause

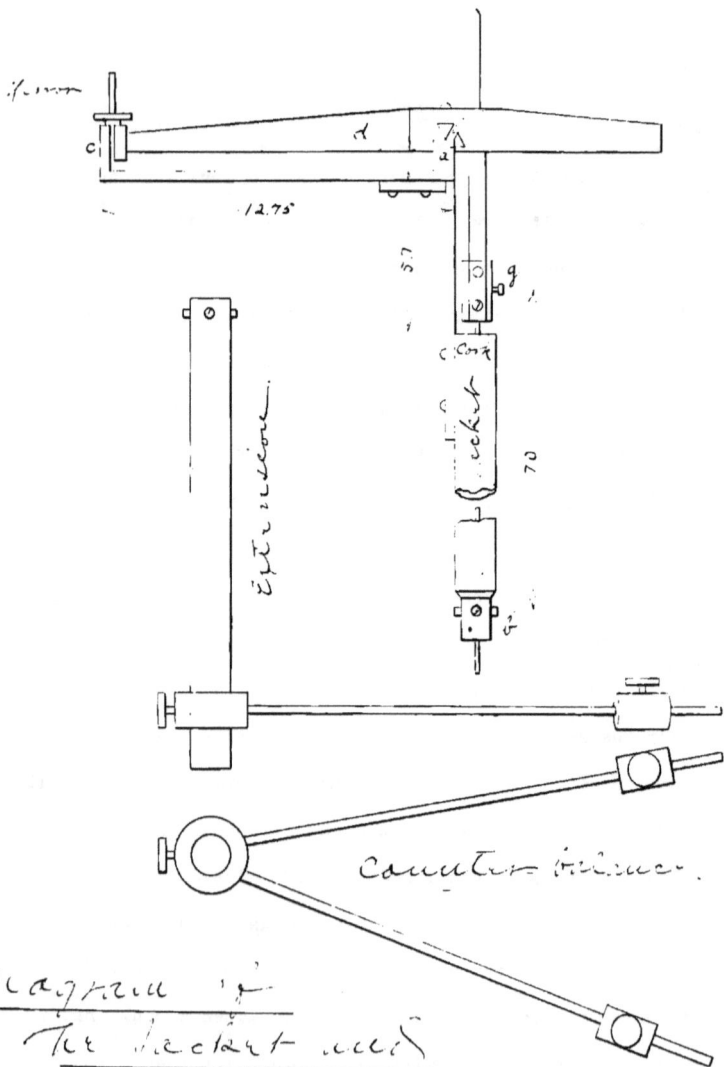

Diagram of
The beckart and
its attachments.

the outer end of the lever to rise or fall. Standing partly
on the end of this lever and partly on the support c is a
little table having three needle-point legs and carrying a
vertical plane mirror. Now any movement of the lever arm
relative to the support c will cause the mirror to tilt; and
by observing with a telescope at some distance the image in
the mirror of a vertical scale the movement of the lever arm
will be greatly multiplied.

If -

L   length of the long arm of the lever

l   length of the short arm of the lever

d   distance from one leg of the brass table to the line
joining the other two legs, resting on lever end.

D   scale distance from the mirror, then the multiplying
power of the apparatus is

$$\frac{L}{l} \times \frac{D \times 2}{d}$$

L   11.672

l - 0.4776

d   0.3335.

D differed slightly for different tests, but was nearly always as great as 170. Thus the multiplying power of the
apparatus was always approximately 25000.

To determine the elongation caused by the tensions used in
most of the tests, and to check the values of the modulus of
elasticity determined from adding small weights, the mirror
at $\underline{c}$ was removed, and another mirror was attached to the lever $\underline{d}$ just above the point $\underline{a}$, so that the apparatus acted as
a simple multiplying lever. This arrangement gave a multi-
plying power of only $2 \frac{D}{1}$, or about 700, and so was capable
of being used for much greater elongations.

    The wire to be tested is always suspended from a
point some distance above the apparatus in such a way that
the jacket has its ends well within the magnetizing solenoid. The solenoid rests upon a support entirely independ-
ent from that of the wire and those parts of the apparatus
attached to the wire, and the two parts are so adjusted that
there is no contact between them.

    I made but two changes in the apparatus itself that
could be supposed to affect the results. In the apparatus
as used by Dr. More and by Dr. Gallaudet, the jacket and its
attachments were not perfectly balanced. The lever $\underline{d}$ was as
nearly balanced as was desirable to have it. But all the
weight of the lever came on one side of the wire; and that
together with the weight of the arm under it, gave quite a
tendency to the jacket to tip sidewise and slightly bend the

wire. Then the amount of this bending would doubtless be
changed by the magnetic stresses; and especially when no
weight was attached to the lower end of the wire, it seemed
that this might possibly have quite an effect upon changes
so small as those to be observed. An adjustable extension
was therefore made to attach to the lower ends of the jacket, extend well down below the magnetizing solenoid, and by
means of sliding weights on its two arms exactly counter-
balance the moment given the jacket by the arm at the top,
together with the lever and mirror. Again, the jacket had
been clamped, at its bottom, to the wire by a single set-
screw. This it seemed might cause a bend in the wire. The
brass plug b was therefore bored out quite a little larger
than the wire. Two extra radial set-screws were put into b,
making three equally spaced ones about the circumference.
Then a little cylinder of brass, slit lengthwise on one side
bored along its axis to fit the wire and just large enough
to slip into the hole in b, was put about the wire at the
point of attachment. In this way by the use of the three
screws the wire could be accurately centered in the jacket.

These precautions seemed necessary for a slight,
almost imperceptible bend in the wire was found by actual
experiment to greatly modify the readings obtained. These

adjustments were so delicate that it was scarcely possible
to judge when they had been properly made except by actual
trial. After Young's modulus had been carefully determined,
it was considered a fair test of the adjustment of the apparatus to put on and off a weight that ought to cause changes
in length approximately equal to those expected to follow
from magnetization. If the computed readings were obtained
the adjustment was assumed to be correct.

A most evident source of error in this work is the
change in length caused by temperature. The greatest value
of $\frac{dl}{l}$ ------ observed in my investigation which seemed to be
due to magnetization was 33.92×10^{-7}. A change of one degree centigrade in the temperature of the wire would give
for $\frac{dl}{l}$ ----- about $120. \times 10^{-7}$. To prevent temperature changes
the magnetizing solenoid was made originally from two coax-
ial brass cylinders in such a way as to leave a space for
water between them. In all the experiments of this investigation water was being continually forced into this space
from the bottom and was over-flowing at the top. Thus there
was a jacket of continuously changing water between the magnetizing coils and the jacket attached to the wire. Yet if
the larger currents were left on even for a short time the
effects of temperature changes were quickly visible. If
the conditions have become steady

and all the apparatus has attained a constant temperature,
and then then a current is put on and allowed to remain, the
resulting higher temperature first affects and expands the
jacket on the wire causing an apparent contraction of the
wire itself, and as the higher temperature reaches the wire
expansion will be observed until at last, in the steady
state resulting, the wire will have apparently expanded or
contracted according to the relative coefficients of heat
expansion for the wire and for the jacket. Fortunately,
these changes are comparatively slow and can be separated
with a limited degree of accuracy from readings that can be
taken quickly. But if one were to take a series of readings
for the modulus of elasticity, using small changes of weight
while the magnetizing current remained on, it might be fore-
seen that the results would apparently vary first in one direction and then in the other, due to the unequal changes in
the length of the jacket and the wire.

Incident to the great multiplying power of the apparatus,
the slightest mechanical vibration of the wire under test
made it is absolutely impossible to read the changes of
length accurately, and, in many cases, to read the figures
of the scale at all. To get a support for the wire as free
from vibrations as possible, a weight of some seventy-five

or eighty pounds was made by filling a wooden box with
bricks and old storage-battery plates and this was suspended
by a single coiled brass spring to an arm projecting from a
sidewall of the room some ten feet above the apparatus. The
spring was so designed that, with the weight it carried, its
period of vibration was very slow, and when the wire to be
tested was suspended from the bottom of the weight, very
little trace of the ordinary vibrations of the building ever
reached the wire. But to have the system work perfectly, it
was necessary that the wire and the weight which it carried
be attached exactly under the centre of mass of the large
weight above. To facilitate this adjustment four projecting
arms were attached to the bottom of the box forming the
weight and on these arms were hung small movable baskets of
shot. When all was properly adjusted, the scale image in
the telescope would be blurred for a half-minute by a wagon
passing on the cobble-stone pavement some twenty or thirty
feet from the apparatus, but otherwise was undisturbed.

GENERAL METHOD OF WORKING.

It was decided that for each kind of iron tested series of elongation readings should be taken for several different tensions, that the elongations should be observed at
each point both with the magnetizing field on and with the
field off, and that for all observations of elongation the
corresponding inductions in the iron should be obtained.

When the wire had been properly adjusted in the apparatus, with a weight attached to the lower end of the wire
to give the desired tension and after the test with the
small weight already mentioned had been made, the first
elongation reading was taken by sending a very weak current
through the magnetizing solenoid, quickly observing the
scale reading in the telescope and then breaking the current
and reading the telescope again. This procedure was repeat-
ed with gradually increasing values of the current until the
maximum current was reached. The elongation readings obtained with the field on, I shall call the "total" elongation,
since it represents both the temporary and the permanent or
residual change in length. The elongation observed after a
certain magnetizing field has been put on and then taken
off, will be called the "residual" elongation for that definite field.

All the currents used in magnetizing were left on
only long enough to get the required readings accurately.
A very careful watch was kept for temperature changes and
when they began to appear, the readings were stopped until
all had cooled down to the steady state again. The currents
used were measured by carefully tested Weston instruments of
different capacity according to the currents to be read.

The elongation data given in this investigation were
all taken the first time the wire was magnetized after it
had been put into the apparatus. No method seemed efficient
in completely demagnetizing the wire while it was in place
in the apparatus. Alternating currents would not do it,
probably because the ends of the wire extended so far beyond
the solenoid. After using the alternating current, the ends
of the wire would still show a decided polarity of the same
kind that it showed before, and the effects upon the elongation curve were very marked. If the magnetizing field was
applied again in the same direction as before the curve was
very similar to the one obtained at first though the amount
of the elongation was somewhat reduced. But if the field
was applied in the opposite direction the first effect with
weak fields was a contraction, followed by a decrease of the
contraction, then by elongation and the rest of the curve

was like other curves for the same wire.

The inductions corresponding to the elongations were obtained from a separate series of readings by means of the ballistic galvanometer, using the method of increasing reversals and breaking the current to get the difference between the total and residual inductions. It was exactly the method used by Professor Rowland [1] in his well known ring experiments. Though actual test had shown the field to be practically uniform along that part of the solenoid where the elongation of the wire was measured, yet, as a precaution against irregularities in the induction of the wire, the test coil was distributed in four equal parts equally spaced along the portion of the wire under observation. For thelower inductions seven hundred turns were used, while for the higher inductions this number was reduced to four hundred turns. The galvanometer with its circuit was calibrated for each connection; and, moreover, the ratio of the deflections obtained by the two different arrangements was always taken each time the change was made.

With soft iron wires the alternate current was applied after the elongation readings had been taken and the induction readings were then made from the same wire. While the elongation curve would have been quite different under the second magnetization, yet actual test showed that the

[1] Rowland, Phil.Mag. (4) vol.46, p.140.

WORK DONE.

I have tested piano-wire in its natural condition under under two different stresses; annealed piano-wire under three different stresses,and soft annealed iron wire under five different stresses, making ten series in all.

The lower tension used for natural piano wire was the smallest that would apparently free the wire from bends, for it could not be obtained in a perfectly straight form. The larger tension was a little more than half way to the elastic limit. These wires have a diameter of 1.25 mm. The same wires used in these tests were then heated to a bright red by passing a current through them and were allowed to cool slowly in the air. The slightly burned outside of the wires was then carefully removed by the use of fine flint-paper after which they were tested under the same tension as before. Since the wires came out of the annealing process perfectly straight it was possible with these wires

to make a test under the tension caused by the apparatus
alone.

But neither with the natural nor the annealed piano-
wire were the changes caused by magnetization very great or
the effects of a permanent strain, especially marked. The
The most extensive investigation was therefore made upon
very soft annealed iron wire. The different pieces were cut
from one continuous piece, and parts taken out at different
places were analyzed by Mr. Nakaseko of this University ,
and were pronounced by him to be very pure and quite free
from carbon as well as from all other impurities. All of
these wires were heated to the bright red state just as in
the case of the annealed piano-wire, but here it was only
to free them from kinks. They gave the same magnetization
curve before and after this process; and while the same elongation curve could not be obtained in the two cases, all of
my work would lead me to think that the difference was due
solely to the bends in the wire before the heating process.

This wire as tested had a diameter of 1.31 mm. Its
behavior was investigated under five different tensions,
ranging from that due to the apparatus only up to the elastic limit of the wire. Under the largest tension the wire

would elongate the usual amount upon the addition of a very small weight, but would require quite a little time to recover when this weight was removed. Possibly because the elastic changes were so slow under this tension, it was impossible to take both the total and residual elongation curves. The best that could be done was to keep the current on all the time and correct as well as could be done for the temperature changes to obtain the curve for total elongation.

Not only were more tests made on this kind of iron, but the readings obtained have been worked out, somewhat more fully. Elongation curves from these results have been plotted both to H, and to I, and in both cases the contractions that Professor Rowland believes will be caused by the force $\dfrac{B^2}{8\pi}$ have been plotted in a separate curve and the elongation curve as modified by this contraction has also been given. For the residual curves this supposed contraction is computed from $\dfrac{(4\pi I)^2}{8\pi}$, the residual value of I , of course, being taken. All these modified curves are drawn on the plates in red while the curves from the observed readings are in black.

The value of Young's modulus used in computing the contraction curves for this iron was 2.12 and was determined both from small changes of tension, using the higher mul-

tiplying power and from larger changes of tension using the
lower multiplying power. The latter method was considered
the more reliable especially in determining the elongations
due to large tensions. But this method could not be used
to investigate the possibility of a change in modulus caused
by magnetization, since it required so much time to adjust
the weights that temperature effects from the magnetizing
current would surely modify the results. By using the smaller changes of tension, no variation in the modulus could be
detected at any stage of magnetization, except such as were
within the limits of error of the method.

It was desired to take the induction readings for
exactly the same values of the magnetizing current that had
been used in the elongation tests; but on account of the bad
condition of the storage batteries that I had to use during
most of my work, I could only make the currents approximately equal in the two cases. The points on each curve, especially where there could be any doubt as to the exact loca-
tion of that part of the curve, were taken very close together; and then through interpolation on the curves, the
corresponding values of elongation and induction could be
found with as great accuracy as the curves could be plotted.

General Results.

Speaking very generally all the curves of elongation when plotted to H, show either no change of length up to nearly the field at which the magnetic saturation occurs or up to that point their elongation curve is similar in shape to the induction curve. Not far beyond this the residual curves become practically horizontal, and show no further change in length of any consequence; while the curve of total elongation begins at nearly the same point to descend along an approximately straight line. The residual curves for soft iron seem to fall quite a little after the higher fields have been applied. But while there seems to be no doubt that a slight shortening of the wire occurs, due probably to a re-arrangement of the molecules under the strain of the higher fields, yet, I am sure that this shortening is exaggerated in my readings by temperature effects that could not be avoided. All the changes of length due to any change in the magnetic condition are very quick; and the almost instantaneous throw of the lever and mirror from one position to another, when quite large, as it was here, caused vibrations that required a little extra time to die out, and consequently allowed some heating of the jacket to occur. All

attempts to damp this vibration involved the possibility of
displacing parts that must move so freely as the lever and
the mirror; and to put the current on and off gradually involved about the same possibility of heating as did the time
required for the vibrations to cease. It thus seemed best
to make these readings as well as could be done and then
give them with the above caution. In some cases I have
drawn the ends of the curves somewhat above the plotted
points, thinking them much more nearly correct there than if
drawn through the points. For the same reason I think it
probable that the ends of the total elongation curves should
show a more marked tendency to curve upwards from the
straight line, as has been found by Bidwell.[1]

Effect of the Field.

In any case the residual curve shows that for quite
a distance beyond the point of saturation no permanent
change of length occurs; and the curve of total elongation
shows, through these same field values, a contraction directly proportional to the field strength. Thus these two

[1] Bidwell, *loc. cit*.

curves, at least along this portion, seem to show clearly
that there is an elongation due directly to the induction in
the iron and a contraction caused by the field, directly
proportional in value to the field strength, as has been suggested by Dr. Gallaudet. Every set of curves given in this
[1]
investigation seems to show a contraction due to the magnetizing field. Where plotted to I., the elongation, in any
one test, is always greater for the same value of induction
when the field is off than when it is on and the curves mod-
ified by the supposed contraction due to $\dfrac{B^2}{8\pi}$ and $\dfrac{(4\pi I)^2}{8\pi}$,
both when plotted to H, and when plotted to I, indicate the
same thing.

Increased tension on the wire apparently causes the
contraction to begin at a lower field and at a lower induction in the iron; but when the elongation reaches the straight
line portion of the total elongation curve plotted to H, it
continues on the same slope without any regard to the tension upon the wire. In other words the straight portions of
all these curves for the same kind of wire are parallel.
For another kind of wire the slope of this portion of the
curve will be entirely different; but here again the curves
for this wire, taken under different tensions, will have
parallel parts. This would seem to indicate that the con-
(1) Gallaudet, loc. cit.

traction caused by the field is not only proportional to
the field but that it depends also upon some definite constant for each kind of iron. For the three different
kinds of iron that I have studied these constants would
have to be about in the ratio of 12.5, 18 and 29 for the
natural piano wire, annealed piano-wire, and the soft iron
respectively. It does not seem possible to identify these
values directly with either the elasticity or the magnetic permeability of the wires.

The curves as modified by the $\frac{B^2}{8\pi}$ correction would
have a slightly smaller absolute value for these constants
but the ratio of their values would be very nearly the
same. Looking at the total elongation curves plotted to
I, it will be seen that the retraction of length does not
appear until the value of μ becomes very much reduced, or
until a relatively large increase of the field is necessary for a small increase of induction in the wire. As the
field necessary to give a certain increase of induction
becomes larger and larger, the curves show a more and more
rapid contraction. Since this part of the curve gradually
becomes more and more nearly vertical it seems evident
that it must approach asymptotically the vertical line
drawn through the maximum or limiting value of I, which

for this soft iron would doubtless be about 1700. This [1]
would indicate that there is no limit to the contraction
which is caused by the field, but the contractions resulting from a definite amount of field increase might finally
become less and less, as Bidwell [2] has observed.

Effects of Tension.

As to the effects of tension, it will be seen that
up to the limit of my magnitizing field, the numerical
sum of the elongation and contraction is nearly the same for all tests of
the same kind of wire regardless of the tension used. But
the greater the tension, the less will be the elongation
and the greater the contraction obtained. Whether we examine the curves as plotted to H or to I, we will see
that, after the contraction begins, the length of the wire
as compared to its length before magnetization, becomes
less as the tension is greater. That is, if we disregard,
for the moment, the elongations caused by the tensions
used, and consider the wires to all have equal lengths
when magnetization begins, then we may say that on the

----

(1) Ewing, Magnetic Ind. p.145.
(2) Bidwell, loc. cit.

parts of the curves, which we are considering, for equal
fields or for equal inductions, the greater the tension
the shorter the wire. For example, if we examine the total elongation curves, as plotted to $\underline{H}$, for soft wire under least and under greatest tension we will see that, after the curves begin to descend, all points on the curve
of greatest tension are about 40 units below those of least
tension. But before any change of length due to magnetization began the elongation caused by the tension, ex-
pressed in the same units used in the curves was 7950 in
one case and only 223 in the other case. What probably
occurs is a reduction of the original elongation of 7950
units by 40 units, and beyond this all the phenomenon is
very nearly as in a case of no tension. Suppose, then,
the wire to become less elastic with magnetization, and
that the value of Young's modulus has increased by about
(1)
one-half per cent   when   the turning point in the magnetization curve is reached; then the elongation that
originally existed will be diminished by one-half per
cent, and we have explained the greater contraction resulting from magnetization when a wire is under tension than
when it has no tension. This change in the modulus of

----

(1)   Bock, loc. cit.

elasticity will explain the reduced elongations of all
other curves in the soft iron series; and a similar ex-
planation will apply to all curves in this investigation,
though the change of Young's modulus must be considerably
less than one-half per cent for the annealed piano-wire
and still less for the natural piano-wire.

In closing, I desire to acknowledge my indebtedness
to Professor Rowland, not only for suggesting this investigation to me, but also for help and kind consideration at
all stages of the work.

I am also greatly indebted to Dr. Ames, to Dr. Dun-
can and to Mr. H. S. Hering for valuable suggestions on many
portions of the experimental work and for a very considerate
interest in the whole investigation.

TABLE I.

Elongation data for piano wire in natural state and under tension of 658 kg. per square cm.

$$\frac{d\,l}{l} \times 10^7 \text{ due to this tension} \quad 2933.$$

H	$-\frac{d\,l}{l} \times 10^7$ (total)	(residual)	H	$\frac{d\,l}{l} \times 10^7$ (total)	(residual)
.914			69.50	- 5.04	3.36
1.83	.00	.00	79.50	- 6.16	3.36
2.74	.00	.00	94.15	- 7.28	3.36
4.11	.00	.00	104.2	- 8.67	3.36
5.49	.00	.00	116.1	-10.09	3.36
8.23	.00	.00	132.7	-11.78	3.36
9.14	.00	.00	143.1	-13.15	3.36
13.09	.00	.56	155.5	-15.12	3.36
15.90	.00	1.12	167.3	-16.24	3.36
19.68	.00	1.68	183.0	-17.92	3.36
22.41	.00	2.24	203.5	-20.73	3.36
26.85	.00	2.80	219.6	- 21.85	3.36
29.95	-0.56	3.36	240.0	-24.66	3.36
33.15	- .84	3.36	260.6	-26.90	3.36
38.10	-1.40	3.36	278.8	- 29.68	3.36
45.08	- 2.52	3.36	301.8	- 31.39	3.36
48.72	- 3.08	3.36	338.5	- 35.30	3.36
53.50	- 3.47	3.36			
59.49	- 3.92	3.36			

TABLE II.

Magnetization data for wire in table I.

H	B	μ	I(total)	I(residual)
1.83	98	53.6	7.7	0
2.75	169	61.5	13.2	.75
6.40	459	71.7	36.1	6.4
8.09	707	87.4	55.6	14.1
10.75	982	91.4	77.3	25.1
12.71	1374	108.1	108.3	36.1
15.50	2394	154.6	189.2	99.2
19.45	5814	299.5	460.5	339.5
21.93	7747	353.0	615.2	483.0
26.65	10167	381.5	807.5	652.5
29.52	11020	374.0	874.5	709.0
32.82	11843	361.2	939.5	751.0
37.65	12578	334.0	999.9	794.0
40.73	13091	322.0	1038.	805.
44.45	13464	303.0	1069.	830.
48.00	13798	287.3	1095.	839.
52.75	14133	268.1	1120.	861.
58.35	14608	250.5	1159.	864.5
66.65	15117	227.0	1199.	876.
76.50	15637	204.5	1238.	883.
91.42	16141	177.0	1238.	890.
101.2	16351	161.5	1294.	896.
112.5	16663	148.2	1318	898
128.4	16848	131.2	1331	900
139.2	16969	121.9	1338	900
151.0	17271	114.4	1363	902
165.0	17345	105.2	1368	902
178.2	17528	98.4	1381	904
198.3	17598	88.8	1385	904
217.4	17937	82.4	1412	906
238.0	18166	76.4	1428	908
256.2	18206	71.1	1428	908
274.5	18375	66.9	1440	910
297.3	18447	62.2	1445	910
331.8	18582	56.0	1453	913

TABLE III.

Elongation data for piano-wire in natural state and under tension of 1949 kg. per sq. cm.

$\dfrac{d\,l}{l} \times 10^7$ due to this tension   3872.

H	$\dfrac{d\,l}{l} \times 10^7$ (total)	(residual)	H	$\dfrac{d\,l}{l} \times 10^7$ (total)	(residual)
.914	.00	.00	53.01	-3.99	2.28
1.83	.00	.00	58.85	-4.56	2.85
2.75	.00	.00	68.98	-5.41	2.85
3.38	.00	.00	78.96	-6.27	2.85
4.12	.00	.00	93.62	-7.74	2.85
4.58	.00	.00	103.8	-9.12	2.85
5.49	.00	.00	115.6	-10.25	2.85
6.41	.00	.00	132.8	12.53	2.85
8.23	.00	.00	142.6	-13.66	3.13
9.14	.00	.00	155.4	-14.80	3.13
10.82	.00	.23	166.9	-15.95	3.42
15.54	.00	1.25	182.9	-18.25	3.42
22.61	.00	1.71	203.8	-20.50	3.42
27.00	-0.57	2.28	219.5	-21.65	3.42
29.73	-1.14	2.28	240.0	25.05	3.42
33.18	-1.71	2.28	256.0	26.78	3.42
37.95	-1.99	2.28	274.4	-30.20	3.42
41.19	-2.39	2.28	302.0	-34.20	3.42
44.85	-2.85	2.28	338.2	39.08	3.42
48.20	-3.42	2.28			

TABLE IV.

Magnetization data for wire in table III.

H	B	μ	I(total)	I(residual)
2.75	153	55.7	12.0	0
3.66	215	58.8	16.8	0
4.71	300	63.7	23.5	4.9
5.49	383	69.8	30.1	8.1
6.54	481	73.5	32.7	12.3
8.23	694	84.4	54.6	13.4
9.23	901	97.5	71.0	29.9
10.88	1213	111.7	95.8	36.6
12.80	1893	147.9	149.5	78.4
15.54	3706	238.1	293.9	185.8
19.65	7908	401.2	562.8	441.8
21.64	8832	407.8	701.8	578.
26.61	10967	412.5	872.2	714.
29.45	11809	401.5	932.5	775.
32.65	12533	383.9	995.5	826
37.25	13277	357.0	1055	879
40.38	13590	336.1	1087	882
43.83	13934	318.1	1105	903
47.40	14267	301.5	1131	928
51.99	14512	279.7	1152	937
57.65	14858	258.0	1179	942
67.65	15288	226.0	1212	956
77.32	15677	202.5	1242	972
91.45	15941	174.5	1263	980
100.5	16201	161.3	1281	984
113.0	16313	144.5	1290	990
128.7	16669	129.6	1313	995
139.5	16750	120.0	1324	998
153.0	17033	111.4	1345	999
163.0	17213	105.8	1357	1000
179.3	17379	97.0	1369	1006
199.0	17599	88.5	1385	1008
215.0	17735	82.5	1395	1010
233.0	17953	77.0	1410	1010
251.8	18052	71.8	1418	1015
274.3	18124	66.2	1422	1020
297.2	18467	62.2	1448	1028
334.0	18884	56.5	1478	1035

TABLE V.

Elongation data for annealed piano-wire under tension

of 62 kg. per sq. cm. (apparatus only).

$\dfrac{d\,l}{l} \times 10^7$ due to this tension   282.

H	$\dfrac{d\,l}{l} \times 10^7$ (total)	(residual)	H	$\dfrac{d\,l}{l} \times 10^7$ (total)	(residual)
.914	.00	.00	58.10	1.84	6.71
1.83	.00	.00	69.32	1.01	6.71
2.74	.00	.00	77.75	.28	6.71
3.66	.00	.00	92.20	-1.68	6.71
5.49	.00	.00	102.35	-3.35	6.71
8.33	.00	.00	114.00	-5.03	6.71
9.15	.28	.56	130.36	-7.55	6.71
10.89	.56	1.01	140.85	-9.50	6.71
12.81	1.01	1.68	153.20	-11.72	6.71
15.54	1.39	2.24	165.65	-13.95	6.71
19.56	1.96	3.35	181.85	-16.75	6.71
22.39	2.24	3.91	202.10	-20.12	6.71
26.70	2.51	4.47	215.9	-22.62	6.71
29.50	2.71	5.03	233.2	-25.12	6.71
32.48	2.71	5.30	256.0	-28.50	6.71
37.27	2.71	5.59	274.5	-31.58	6.71
40.22	2.71	5.87	301.9	-35.75	6.71
43.95	2.51	6.15	338.2	-40.20	6.71
47.16	2.35	6.44			
52.22	1.96	6.71			

TABLE VI.

Magnetization data for wire in table V.

H	B	μ	I(total)	I(residual)
1.83	102.	55.7	7.9	0
2.74	187.	68.3	14.7	1.6
3.75	270.	72.0	21.2	3.3
4.72	379	80.4	29.8	4.5
8.33	843	101.1	66.4	22.2
9.28	1014	109.3	80.1	23.4
10.99	1384	126.1	109.2	43.7
12.90	1998	154.9	158.0	69.7
15.70	3586	228.3	284.0	179.9
19.84	5995	302.1	476.0	347.0
22.65	7099	314.0	564.0	424.5
27.00	8452	313.0	672.0	506.5
29.60	9130	309.0	724.0	544.5
32.59	9723	299.0	771.0	573.0
37.50	10598	283.0	841.0	627.0
40.65	10961	270.0	869.0	642.5
44.28	11464	259.0	909.0	655.0
47.63	11778	247.0	934.0	671.8
52.78	12433	236.0	986.0	698.0
58.56	12779	217.5	1013.	718.5
69.10	13449	194.9	1065	725.0
78.25	13898	177.6	1100	736.
93.72	14574	155.6	1152	756
103.4	15053	145.7	1190	758
115.3	15605	135.5	1232	760
131.3	15671	119.4	1238	761
141.7	15882	112.1	1253	763
154.6	16155	104.5	1274	763
166.9	16527	99.0	1302	766
183.0	16683	91.3	1313.	768
203.8	16964	83.3	1333	768
219.5	17170	78.3	1349	770
240.1	17260	71.9	1355	774
258.2	17468	67.6	1369	776
279.1	17649	63.4	1382	780
304.1	17844	58.4	1398	780
338.1	18138	53.6	1417	785

TABLE VII.

Elongation data for annealed piano-wire under tension
of 699 kg. per sq. cm.

$$\frac{d\,l}{l} \times 10^7 \text{ due to this tension} \quad 3178.$$

H	$\dfrac{d\,l}{l} \times 10^7$ (total)	(residual)	H	$\dfrac{d\,l}{l} \times 10^7$ (total)	(residual)
.914	.00	.00	52.18	-0.567	5.39
1.83	.00	.00	58.10	-1.42	5.39
2.74	.00	.00	69.45	-3.42	5.39
3.66	.00	.00	78.22	-4.54	5.39
4.57	.00	.00	93.20	-7.37	5.39
5.49	.00	.00	103.3	9.06	5.39
8.14	.00	.17	114.7	-10.78	5.39
9.23	.17	.567	131.2	-14.17	5.39
11.95	.567	.850	141.6	-15.88	5.39
12.81	.850	1.42	157.9	-18.70	5.39
15.54	1.70	2.27	166.4	-21.00	5.39
19.69	2.55	4.25	182.4	-23.25	5.39
21.99	2.84	4.82	203.2	-27.20	5.39
27.00	2.84	5.11	219.2	-30.61	5.39
29.72	2.27	5.11	237.9	-32.88	5.39
32.90	1.99	5.39	258.2	-36.84	5.39
37.92	1.70	5.39	274.2	-40.26	5.39
40.98	1.13	5.39	304.1	-44.22	5.39
44.38	.567	5.39	340.8	-49.02	5.39
48.00	.00	5.39			

Magnetization data for wire in table VII.

H	B	μ	I(total)	I(residual)
1.83	99	55	7.7	0
2.65	176	66.4	13.8	2.5
3.66	269	73.5	21.1	3.0
4.67	362	77.6	28.4	3.1
5.49	433	79.0	33.9	6.7
8.09	837	103.6	65.9	25.7
9.15	1024	112.0	80.8	27.7
10.74	1399	130.5	110.6	47.2
12.66	2077	164.5	164.2	88.3
15.18	3587	236.0	284.0	182.0
16.91	4352	258.0	345.0	227.0
19.19	5689	295.6	450.5	324.0
20.80	6306	303.8	500.0	361.8
22.41	6952	310.0	551.0	409.2
30.49	8990	294.0	715.5	553.5
36.80	10327	281.0	819.0	608.8
39.75	10740	271.0	852.5	629.0
43.25	11143	258.0	883.0	651.0
47.18	11577	245.5	917.5	656.0
51.45	11972	233.0	949.0	668.0
56.95	12407	218.3	984.0	679.0
69.01	13369	193.8	1060.	690.
77.00	13707	178.0	1085	695
91.45	14482	158.3	1142	700
101.6	14592	143.7	1152	706
112.9	15043	133.0	1190	712
129.0	15389	119.2	1214	712
139.3	15629	112.3	1232	714
151.4	15871	104.6	1251	716
166.2	16066	96.4	1266	717
178.2	16428	92.2	1295	718
198.1	16558	83.6	1301	720
215.4	16915	78.5	1329	722
235.4	17055	72.6	1339	724
251.5	17282	68.7	1356	726
274.5	17375	63.2	1361	726
297.2	17497	58.9	1369	730
327.0	17917	54.8	1399	733

TABLE IX.

Elongation data for annealed piano-wire under tension
of 1949 kg. per sq. cm.

$$\frac{d\,l}{l} \times 10^7 \text{ due to this tension} \quad 8872.$$

H	$\dfrac{d\,l}{l} \times 10^7$ (total)	(residual)	H	$\dfrac{d\,l}{l} \times 10^7$ (total)	(residual)
.914	.00	.00	48.00	- 1.395	4.75
1.83	.00	.00	52.50	- 1.95	4.75
2.74	.00	.00	58.49	- 2.79	4.75
3.66	.00	.00	68.00	- 4.29	4.75
4.57	.00	.00	77.62	-5.75	4.75
5.49	.223	.223	92.45	- 8.53	4.75
6.40	.446	.558	102.8	- 9.93	4.75
8.13	.837	1.114	114.3	-11.71	4.75
9.14	1.40	1.67	130.3	-14.50	4.75
10.74	2.01	2.23	140.7	-16.17	4.75
12.72	2.34	2.79	153.2	-18.38	4.75
15.53	2.68	3.68	164.5	-20.07	4.75
19.49	2.79	4.35	180.5	-23.41	4.75
22.13	2.23	4.46	200.6	26.21	4.75
27.55	1.95	4.75	216.8	- 29.00	4.75
28.80	1.40	4.75	235.7	-31.78	4.75
32.02	.837	4.75	256.0	35.70	4.75
36.55	.279	4.75	274.5	-37.90	4.75
40.68	-0.279	4.75	297.5	42.42	4.75
44.76	- .837	4.75	331.5	47.45	4.75

TABLE X.

Magnetization data for wire in table IX.

H	B	μ	I(total)	I(residual)
1.83	113	61.8	8.9	0
2.74	187	68.3	14.6	2.3
3.84	285	74.3	22.3	5.1
5.57	466	83.7	36.6	6.8
6.58	610	92.9	48.0	14.1
8.22	894	108.8	70.5	25.6
9.45	1119	118.2	88.3	30.6
10.87	1524	140.5	120.4	58.0
12.88	2279	177.0	180.3	91.6
15.63	3891	249.2	228.6	196.0
19.82	6230	314.5	494.5	361.0
21.93	6932	316.6	549.9	403.5
27.08	8527	314.8	675.8	506.2
29.82	9090	304.9	721.0	536.5
33.08	9778	296.1	775.0	573.0
38.02	10528	277.0	835.2	609.5
40.92	10821	264.6	859.0	622.5
44.55	11245	253.0	891.0	640.0
48.00	11548	241.0	914.5	666.0
52.64	11933	226.8	945.0	670.0
58.45	12409	212.1	982.5	683.0
69.00	13119	190.5	1037.	704.
78.00	13598	174.2	1074	725
93.65	14244	152.6	1125	730
103.8	14434	139.2	1139	734
115.2	14695	127.5	1159	736
132.7	15333	115.7	1209	740
154.0	15754	102.2	1239	742
165.5	15846	95.7	1248	744
181.4	16082	88.6	1264	744
201.2	16402	81.6	1288	744
219.5	16819	76.5	1319	746
237.7	16868	70.8	1323	746
256.0	17256	67.4	1353	748
274.3	17284	63.2	1354	748
301.5	17552	58.2	1372	748
338.1	17698	52.4	1379	751

TABLE XI.

Elongation data for soft, annealed iron under tension

of 48.3 kg. per sq. cm. (apparatus only)

$$\frac{d\,l}{l} \times 10^{7} \text{ due to this tension} \quad 223.$$

H	$\frac{dl}{l} \times 10^{7}$ (total)	$\frac{dl}{l} \times 10^{7}$ (residual)	H	$\frac{d\,l}{l} \times 10^{7}$ (total)	$\frac{dl}{l} \times 10^{7}$ (residual)
.549	.000	.00	37.55	33.07	31.70
.914	.285	.285	41.80	32.50	32.20
1.37	.855	.684	45.32	32.05	32.60
1.93	2.05	1.82	48.91	31.62	33.06
2.28	3.82	3.71	53.51	30.88	33.16
2.74	4.85	4.56	58.75	29.74	33.25
3.16	6.39	5.82	64.50	28.50	33.34
3.61	8.55	7.52	73.18	26.78	33.50
4.25	11.97	9.96	87.50	23.36	33.61
4.81	14.53	12.18	96.50	20.62	33.61
5.67	17.84	14.82	106.9	17.88	33.61
6.36	21.64	18.13	114.1	15.95	33.61
8.55	25.63	21.35	121.1	13.75	33.61
9.79	27.38	23.08	130.3	11.40	33.61
11.32	29.20	24.80	140.7	8.98	33.34
13.33	30.78	26.35	148.5	6.82	33.06
16.23	32.50	27.93	166.8	.855	33.06
20.34	33.63	29.64	183.2	- 3.42	32.77
23.3	33.92	30.21	190.5	· 5.13	32.50
30.72	33.63	31.35	204.4	- 9.12	31.92
34.18	33.55	31.65	222.5	14.52	31.06
			251.4	- 22.22	30.20

TABLE XII.

Magnetization data for wire in table XI.

H	B	μ	I (total)	I (residual)	$\frac{2\pi \times 10^7}{8\pi M}$ (total)	(residual)
.586	184.8	269	14.6	0	.006	0
1.05	317.	301	25.1	1.2	.02	.004
1.83	740.	404	58.7	17.3	.10	.008
2.29	1122.	490	89.2	37.5	.24	.04
2.75	2608	913	207.	131.5	1.28	.512
3.20	3823	1195	304	220.2	2.74	1.43
3.68	5089	1382	404	302.5	4.85	2.72
4.30	6449	1502	513.	393.8	7.80	4.58
4.95	7725	1561	614	487.5	11.21	6.97
5.67	9206	1628	732	595.4	15.95	10.48
6.86	10727	1566	854	710	20.50	14.90
8.65	11939	1397	949	794	23.73	18.63
9.69	12610	1302	1004	837	29.90	20.7
12.61	13503	1054	1075	903	34.30	24.1
15.55	14171	912	1126	954	37.75	26.9
19.66	14710	749	1170	974	40.65	28.1
21.02	14831	707	1180	986	41.38	28.7
26.67	15207	571	1209	1003	43.45	29.7
32.83	15483	472	1230	1010	45.10	30.2
37.45	15727	420	1249	1017	46.40	30.6
44.05	16004	364	1271	1025	48.20	31.0
47.50	16098	339	1278	1028	48.6	31.2
57.91	16558	286	1314	1035	51.1	31.6
69.05	16769	243	1329	1037	52.6	31.8
82.35	17012	207	1348	1040	54.4	32.0
91.50	17102	187	1354	1045	54.9	32.3
101.5	17202	169	1361	1056	55.6	32.9
107.9	17358	161	1374	1058	56.5	33.2
114.8	17505	153	1384	1060	57.5	33.3
123.9	17614	142	1392	1060	58.3	33.3
134.9	17915	133	1416	1065	60.3	33.3
146.1	17956	123	1419	1065	60.5	33.6
159.6	18260	114	1441	1065	62.5	33.6
177.1	18497	104	1458	1070	64.2	33.9
197.6	18688	95	1471	1070	65.5	33.9
208.0	18768	90	1477	1070	66.5	33.9

TABLE XIII.

Elongation data for soft annealed iron wire under tension of 228. kg. per sq. cm.

$\dfrac{dl}{l} \times 10^7$ due to this tension 1052.

H	$\dfrac{d\,l}{l} \times 10^7$ (total)	(residual)	H	$\dfrac{d\,l}{l} \times 10^7$ (total)	(residual)
1.01	1.73	1.72	40.20	29.27	32.78
1.37	4.02	3.45	44.42	28.68	32.78
1.83	5.75	5.17	47.65	27.00	32.78
2.29	8.05	7.18	52.25	25.85	32.78
2.74	9.63	9.19	57.95	24.12	32.78
3.11	11.48	10.33	61.58	22.99	32.78
3.56	13.78	12.62	71.00	20.67	32.78
4.25	16.64	14.93	84.78	17.23	33.00
4.94	18.65	17.22	93.65	14.93	33.00
5.62	21.22	19.55	104.0	12.05	33.00
6.76	24.10	21.83	111.0	9.77	32.78
8.50	26.42	24.12	118.4	7.47	32.78
9.59	27.58	25.85	128.0	4.89	32.43
11.20	29.28	27.26	139.4	1.18	32.18
13.16	30.44	28.18	149.9	-1.436	32.18
16.00	31.01	29.86	164.5	-5.168	32.18
20.13	31.59	31.00	182.4	-9.77	31.88
22.52	31.85	31.60	186.3	-10.92	31.88
26.63	31.59	31.87	203.0	-16.09	31.60
29.42	31.32	32.18	217.8	-19.52	31.01
32.70	30.42	32.43	244.6	25.26	29.88
37.48	29.83	32.78	265.6	32.70	27.62

tion data for wire in table XIII.

μ	I(total)	I(Residual)	$\dfrac{B^2 \times 10^?}{8\,\pi\,M}$ Total	Residual
389.	31.1	0.16	.028	
472	51.4	4.2	.078	.008
904	163.9	81.7	.798	.197
1260	270.6	173.0	2.17	.888
1585	398.0	277.	4.70	2.27
1764	507.2	388.	7.64	4.45
1970	759.0	560.	17.14	9.27
1770	968.0	821.	27.8	19.9
1530	1039.	893.	32.0	23.6
1390	1074	930	34.2	25.6
1280	1105	957	36.3	27.1
1075	1133	983	38.1	28.6
904	1167	1010	40.6	30.3
733	1194	1030	42.4	31.4
659	1223	1044	44.4	32.2
562	1245	1062	46.18	33.3
513	1270	1072	47.4	34.0
463	1272	1075	47.9	34.2
434	1275	1078	48.4	34.4
334	1313	1034	51.8	34.5
282	1336	1085	53.3	34.8
253	1348	1086	55.2	34.9
220	1360	1098	55.4	35.6
188	1390	1102	57.6	35.9
171	1400	1106	58.8	36.1
155.6	1414	1106	60.1	36.1
147.0	1425	1106	61.0	36.1
138.7	1432	1106	61.3	36.1
129.0	1445	1106	62.8	36.1
120.0	1463	1106	64.4	36.1
112.0	1472	1106	65.4	36.1
103.4	1489	1106	67.0	36.1
94.6	1502	1106	68.3	36.1
90.2	1505	1106	68.7	36.1
82.4	1509	1106	69.2	36.3
79.4	1520	1108	70.3	36.3
74.4	1548	1110	73.0	36.4

TABLE XV.

Elongation data for soft annealed iron wire under
tension of 430. kg. per sq. cm.

$$\frac{d\ l}{l} \times 10 \text{ due to this tension } 1988.$$

H	$\dfrac{d\ l}{l} \times 10^7$		H	$\dfrac{d\ l}{l} \times 10^7$	
	(total)	(residual)		(total)	(residual)
.457	0.29	0.28	42.99	20.92	25.92
.915	.853	.68	46.48	20.38	26.11
1.51	1.71	1.42	50.68	19.62	26.19
1.82	1.99	1.82	54.82	18.76	26.19
2.42	3.41	2.84	60.08	17.61	26.19
2.88	4.55	4.09	67.00	15.92	26.22
3.43	6.27	5.69	76.50	13.41	26.40
3.93	7.97	7.40	91.95	9.96	26.72
4.62	10.52	9.68	101.4	6.95	26.72
5.26	12.39	11.37	113.2	3.99	26.72
6.12	14.51	13.64	129.5	- 0.57	26.60
7.31	17.06	15.91	139.8	- 3.13	26.60
9.14	19.21	18.18	152.4	- 6.27	26.40
10.40	20.19	19.33	163.7	- 9.68	25.60
12.21	21.32	20.44	178.9	13.93	25.01
14.41	21.71	21.18	198.6	-19.21	24.42
17.54	22.78	22.77	228.5	26.16	24.15
22.18	23.20	23.78	246.6	31.00	23.02
25.32	23.20	24.20	262.5	36.12	21.00
30.18	22.78	24.78	283.2	42.99	18.77
33.48	22.20	25.01	308.3	54.50	15.32
37.39	21.61	25.60			

Magnetization data for wire in table XV.

H	B	μ	I (total)	I(Residual)	B x 10⁷ Total	Residual
					8 ~ M	
1.46	464	318	36.8	8.09	.04	.
1.92	752	392	59.7	23.3	.10	.02
2.29	1496	654	119.0	65.3	.42	.12
2.75	3235	1178	255.2	182.0	1.97	.98
3.39	5658	1672	442.5	347.5	7.00	3.57
4.58	9405	2060	748.5	634.5	16.3	11.9
5.22	10695	2060	852.5	733.8	21.4	15.9
6.08	11616	1910	924.2	820.0	25.3	19.9
7.32	12437	1710	994.	880.0	29.3	22.9
9.04	12909	1430	1025.	915.0	31.3	24.8
10.29	13270	1290	1055	937.5	33.1	26.0
12.09	13572	1120	1079	976.0	34.6	28.2
14.24	13864	973	1102	987.5	36.1	28.9
17.20	14197	826	1128	1008.	37.8	30.1
21.60	14532	673	1154	1026.	39.7	31.1
24.63	14775	600	1173	1042.	41.0	32.1
29.28	14959	511	1187	1053.	42.0	32.3
32.28	15082	468	1198	1053	42.7	32.3
35.81	15296	427	1213	1053	43.9	32.3
38.60	15399	399	1221	1053	44.5	32.3
41.00	15441	377	1225	1055	44.8	32.9
45.00	15605	344	1237	1056	45.8	33.0
52.63	15793	300	1252	1058	46.9	33.1
58.12	15938	274	1263	1058	47.7	33.1
68.60	12229	236	1285	1058	49.5	33.1
78.35	16408	210	1298	1059	50.6	33.1
93.15	16733	179	1324	1070	52.6	33.9
102.9	16903	165	1336	1070	53.7	33.9
114.3	17064	149	1348	1070	54.5	33.9
130.6	17411	133	1374	1074	56.8	34.3
140.0	17462	123	1380	1076	57.3	34.3
154.3	17674	115	1393	1076	58.5	34.3
166.5	17907	107.5	1411	1076	60.3	34.3
183.1	17983	98.0	1416	1078	60.7	34.4
203.8	18244	89.0	1430	1078	62.5	34.4
226.3	18366	81.3	1443	1078	63.3	34.4
249.2	18429	73.9	1446	1080	63.7	34.5
270.0	18790	69.5	1473	1080	66.3	34.5
297.5	18947	63.5	1483	1087	67.4	34.5
341.6	19422	56.8	1516	1092	70.8	35.3

TABLE XVII.

Elongation data for soft annealed iron wire under tension of 752 kg. per sq. cm.

$$\frac{d\,l}{l} \times 10^7 \text{ due to this tension } - 3480.$$

H	$\dfrac{d\,l}{l} \times 10^7$ (total)	(residual)	H	$\dfrac{d\,l}{l} \times 10^7$ (total)	(residual)
0.46	0.40	0.29	58.70	_ 1.60	15.25
1.14	2.40	2.40	69.51	- 4.57	15.30
1.28	2.86	2.86	77.78	- 7.42	15.30
1.33	4.86	5.03	84.52	- 9.14	15.30
2.01	6.17	6.40	92.95	-11.71	15.19
2.38	7.42	7.82	102.4	-14.28	15.14
2.56	7.99	8.57	113.1	-17.25	15.14
2.88	8.85	9.42	130.2	-22.19	14.95
4.57	10.28	11.42	140.1	-25.15	14.28
8.01	10.86	13.02	152.0	-29.17	13.71
10.70	10.17	13.72	159.0	-33.00	13.42
12.62	9.88	13.88	180.6	-37.26	13.13
15.55	9.41	14.29	200.0	-42.73	12.00
19.56	8.56	14.57	205.8	-44.55	11.70
22.40	7.87	14.74	222.0	50.00	10.28
26.72	6.85	14.83	240.0	55.45	9.14
32.95	5.26	15.02	263.0	62.00	8.00
37.76	3.71	15.08	293.0	69.65	6.86
44.50	2.06	15.08			
52.98	.00	15.20			

TABLE XVIII.

Magnetization data for wire in table XVII.

H	B	μ	I(total)	I(Residual)	$B^2 \times 10^7$ $8\pi M$ Total	Residual
1.28	395	309	31.4	5.37	.03	.
2.01	1150	572	91.4	36.0	.24	.04
2.51	2317	921	184.0	105.8	1.01	.33
2.97	3725	1250	296.3	179.8	2.6	.956
3.39	4735	1400	377.0	251.8	4.1	1.87
3.98	6299	1590	501.0	372.6	7.5	4.11
4.58	8130	1780	647.0	502.5	12.4	7.46
5.46	10225	1876	813.3	658.0	19.7	12.8
6.41	11396	1780	908.0	713.0	24.4	15.1
7.91	12278	1560	977.5	845.0	28.3	21.2
10.51	12881	1225	1025.	880.0	31.2	22.9
12.46	13112	1050	1042	889.0	32.3	23.4
15.09	13615	904	1083	915.0	34.9	24.8
18.96	13909	735	1106	935.0	36.3	25.9
21.47	14001	651	1113	962.5	36.8	27.4
25.60	14266	557	1135	963	38.2	27.6
28.37	14558	513	1158	968	39.8	27.7
31.36	14631	467	1162	975	40.2	28.2
36.12	14736	408	1170	977	40.8	28.3
39.10	14789	379	1174	988	41.2	28.9
46.00	14916	324	1183	992	42.8	29.2
50.38	15050	299	1194	1000	43.4	29.6
64.25	15564	242	1234	1004	45.4	29.9
84.60	15865	187	1258	1006	47.2	30.0
92.38	16172	175	1282	1012	49.1	30.4
102.4	16342	160	1294	1016	50.3	30.6
116.6	16397	141	1297	1016	50.6	30.6
125.7	16466	131	1302	1018	51.9	30.7
147.7	16898	114	1335	1020	53.5	30.8
164.2	17144	104.6	1351	1024	55.1	31.0
180.6	17301	95.8	1364	1026	56.1	31.2
212.4	17712	83.4	1394	1028	59.0	31.3
260.5	17981	69.0	1411	1032	60.6	31.5

TABLE XIX.

Elongation data for soft annealed iron wire under tension of 1720. kg. per sq. cm.

$\dfrac{d\,l}{l} \times 10^7$ due to this tension   7950.

H	$\dfrac{d\,l}{l} \times 10^7$ (total)	H	$\dfrac{d\,l}{l} \times 10^7$ (total)
.458	.229	46.00	− 8.94
.905	.671	46.18	−10.06
1.37	1.23	55.76	−11.19
1.83	3.36	66.32	−14.41
2.29	4.47	75.86	−16.76
2.74	5.03	89.78	−20.39
3.38	5.82	99.15	−23.08
3.93	6.26	110.6	−25.98
4.53	6.26	126.6	−30.45
5.21	6.15	135.7	−33.25
6.31	5.36	148.1	−36.42
7.77	4.74	159.1	−40.00
8.87	4.24	175.6	−43.92
12.34	2.52	194.2	−48.65
18.28	−0.56	205.8	−51.95
21.50	−1.67	221.9	−56.05
25.78	−3.13	237.8	−60.09
28.55	−3.91	256.0	− 64.75
31.62	−4.81	283.5	74.55
36.45	−6.05		
39.40	−6.88		
42.55	−7.82		

Magnetization data for wire in table XIX.

H	B	μ	I(total)	$\dfrac{B^2 \times 10^7}{8\pi \times I\ (Total)}$
.914	176	192	13.3	
1.37	313	228	24.3	.02
2.28	570	250	45.2	.06
3.74	709	239	56.3	.09
3.20	914	286	72.5	.13
3.61	1203	334	95.5	.27
4.25	1973	464	156.4	.73
4.85	3325	685	264.1	2.07
5.58	6711	1183	533.6	8.46
6.72	9557	1420	760.	15.3
8.36	11108	1330	985.	23.2
9.56	11570	1212	921	25.2
11.14	12151	1090	966	27.8
13.25	12583	950	1002	29.8
16.10	13086	814	1040	32.1
20.21	13730	680	1092	35.0
23.10	13903	604	1105	36.4
27.65	14258	517	1133	38.2
30.64	14391	470	1144	38.8
33.93	14534	429	1150	39.6
38.98	14739	379	1170	40.5
42.25	14962	355	1198	42.0
45.8	15106	330	1199	42.8
49.6	15160	306	1202	43.2
54.9	15255	278.6	1210	43.6
60.9	15481	255.0	1229	45.0
69.5	15670	226.0	1241	46.1
78.0	16065	206.0	1272	48.4
94.4	16194	171.5	1281	49.2
104.2	16554	159.0	1310	51.3
116.6	16667	143.0	1318	52.0
133.4	17013	127.5	1343	54.3
156.4	17336	111.0	1368	56.5
184.5	17545	95.2	1382	57.9
204.2	17904	88.0	1393	60.1
224.1	17974	80.4	1412	60.5
247.0	18167	73.4	1426	61.9
263.0	18433	70.0	1445	63.7
283.8	18534	65.5	1458	64.6

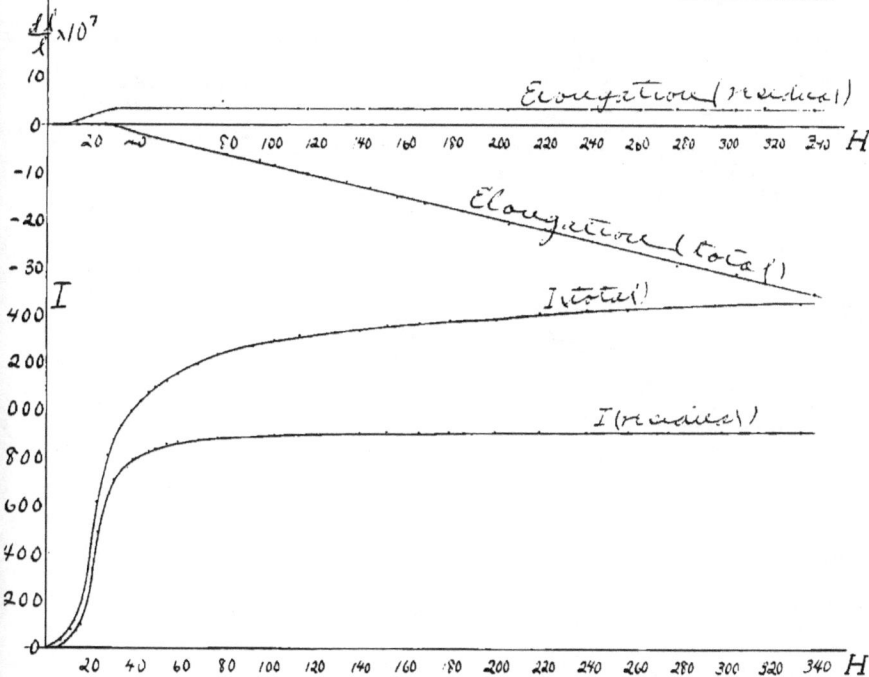

Natural Piano-Wire:
Tension 605 ½ kgr. ... mm.
Data in Tables I and II

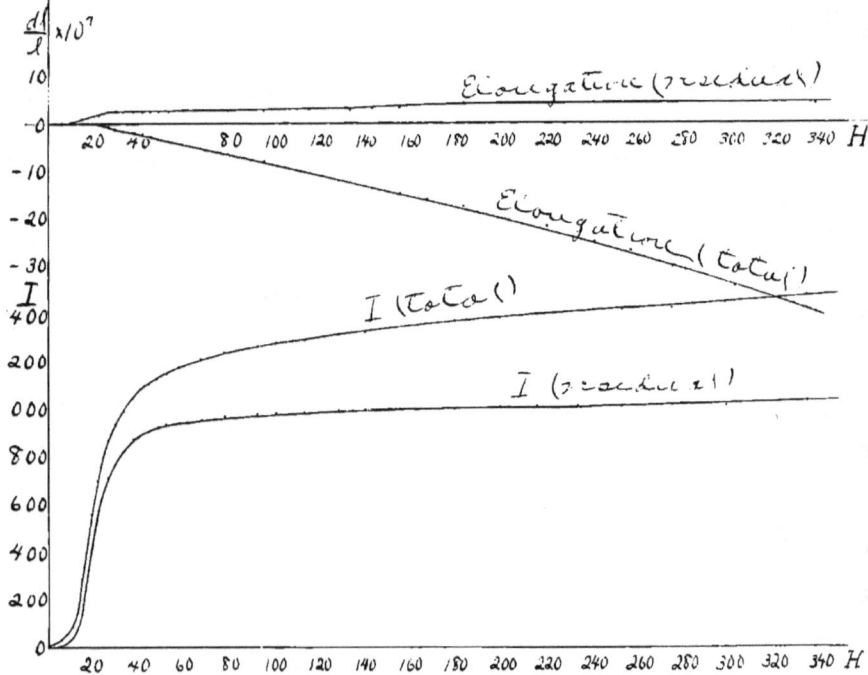

Natural Piano-Wire.
Tension 1949 ... per sq cm.
Data in Tables III and IV

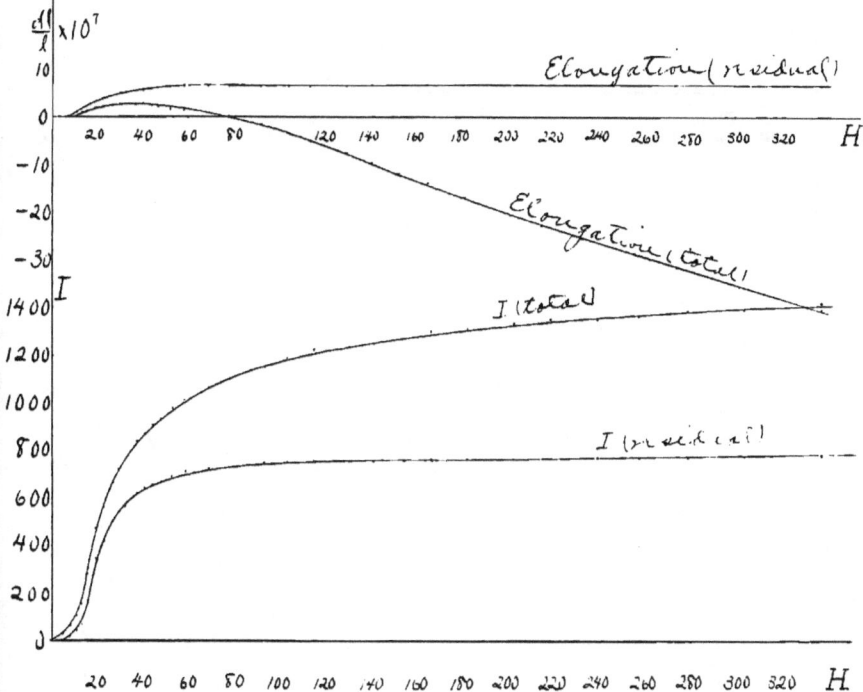

Annealed Piano-Wire.
Tension 0.2 kg per sq. cm
Data in Tables 5 and 6

$\frac{dl}{l} \times 10^7$

Elongation (residual)

Elongation (total)

I (total)

I (residual)

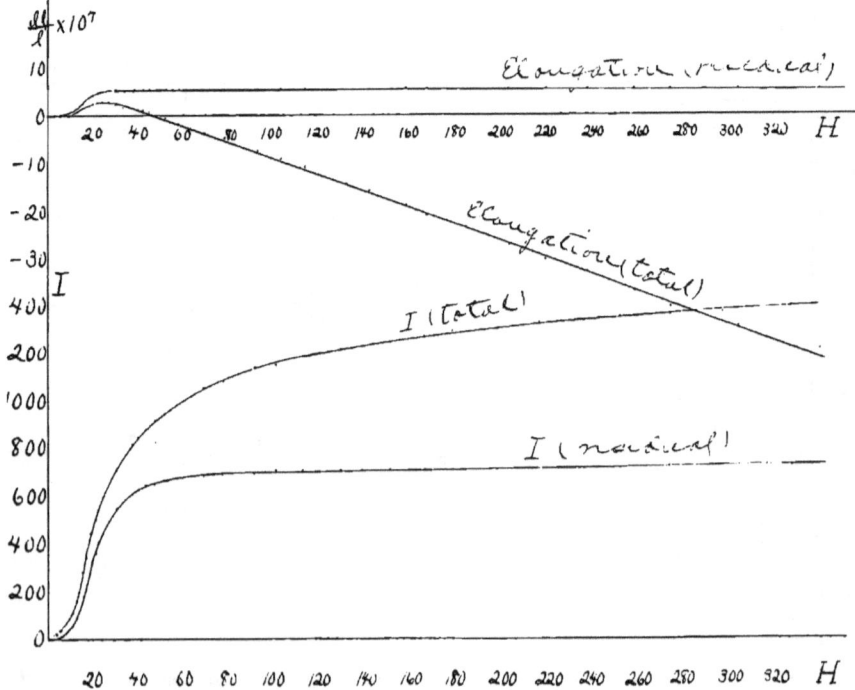

Annealed Piano-Wire

Tension 649 gr. per sq. cm.

Data in Tables VII and VIII

$\frac{\delta l}{l} \times 10^7$

Elongation (residual)

Elongation (total)

I (total)

I (residual)

H

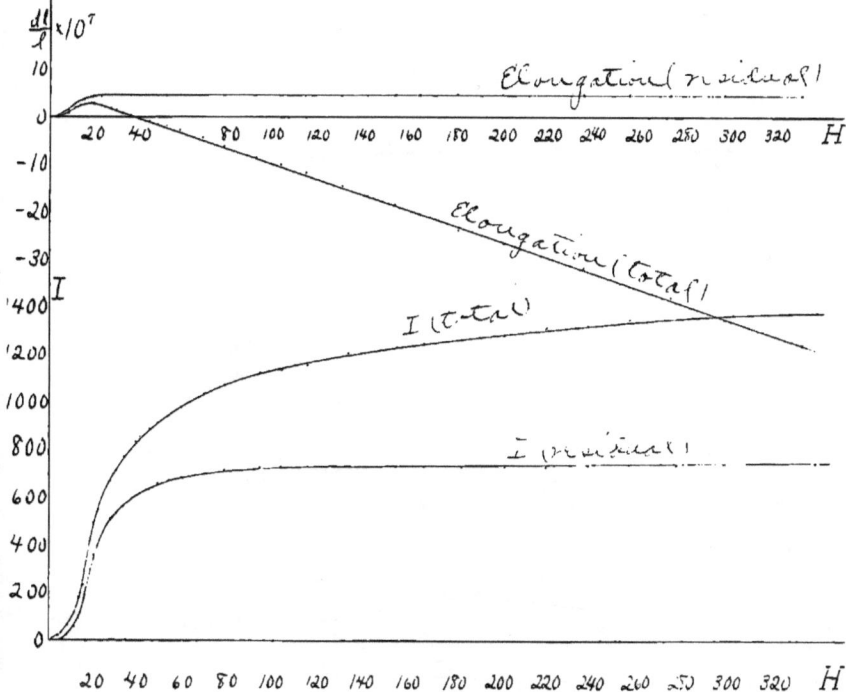

Annealed Piano Wire
Tension 1949 ^{kg} per sq. cm.
Data in Tables I and II

$\frac{dl}{l} \times 10^7$

10

0

−10

−20

−30

Elongation (residual)

20 40 80 100 120 140 160 180 200 220 240 260 280 300 320 H

Elongation (total)

400 I

1200

1000

800

600

400

200

0

I (total)

I (residual)

20 40 60 80 100 120 140 160 180 200 220 240 260 280 300 320 H

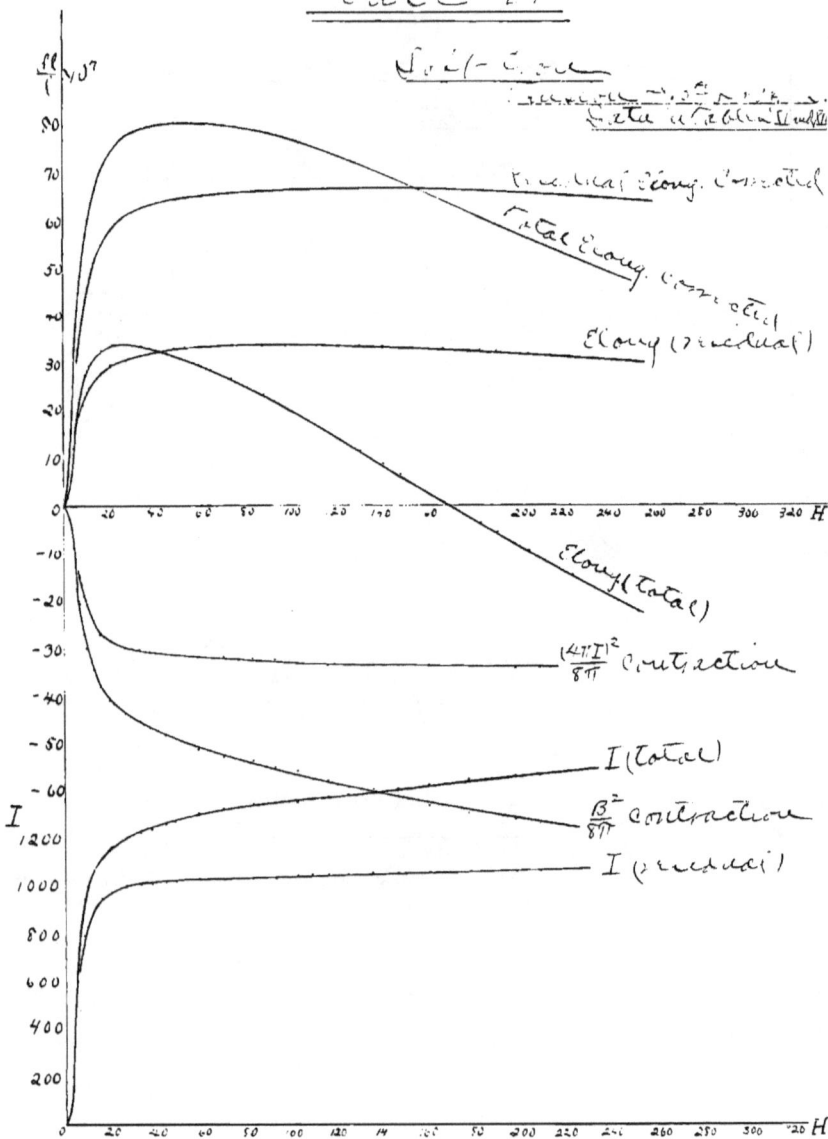

Plate I.

Soft Iron

Residual Elong. Corrected
Total Elong. Corrected
Elong (residual)
Elong (Total)
$\frac{(4\pi I)^2}{8\pi}$ Contraction
I (Total)
$\frac{B^2}{8\pi}$ contraction
I (residual)

Soft Iron

Traction and kg per sq cm
data in Tables XIII and XIV

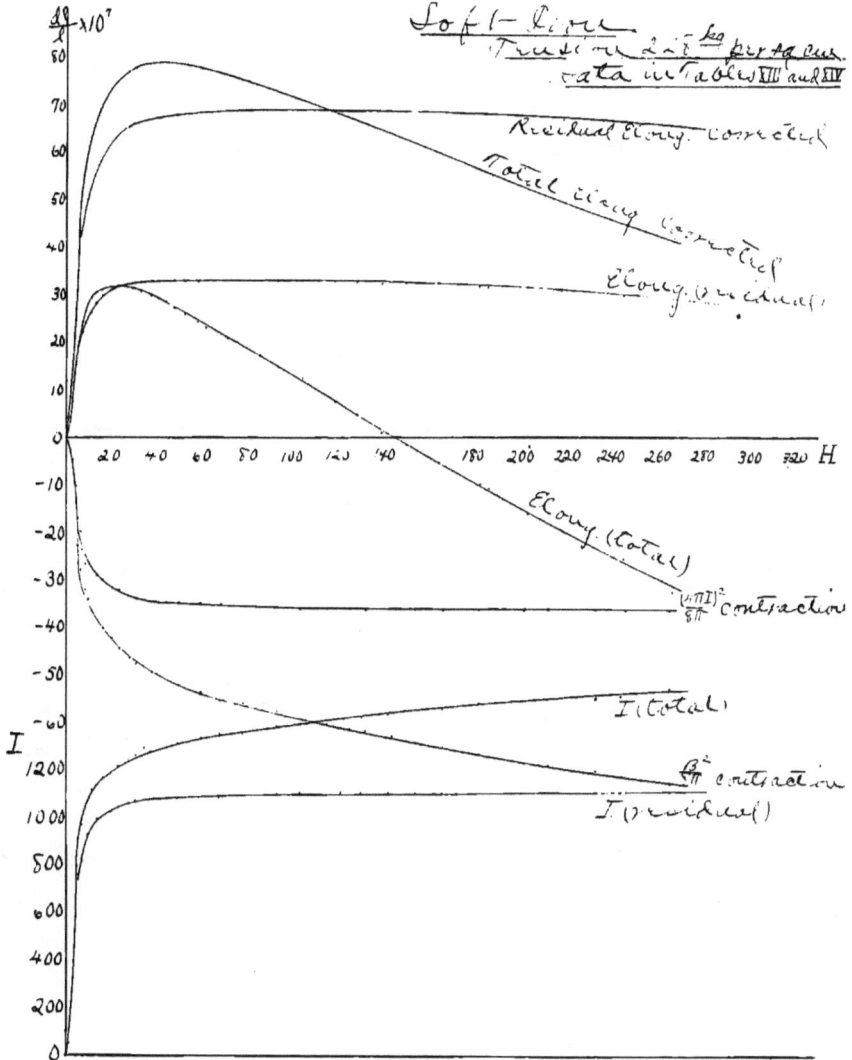

$\frac{d l}{l} \times 10^7$

80
70
60
50
40
30
20
10
0

Residual Elong. corrected

Total elong. corrected

Elong. (residual)

20 40 60 80 100 120 140 180 200 220 240 260 280 300 320 H

-10
-20
-30
-40
-50
-60

Elong. (total)

$\frac{(\pi I)^2}{8\pi}$ contraction

I (total)

$\frac{I^2}{8\pi}$ contraction

I (residual)

I

1200
1000
800
600
400
200
0

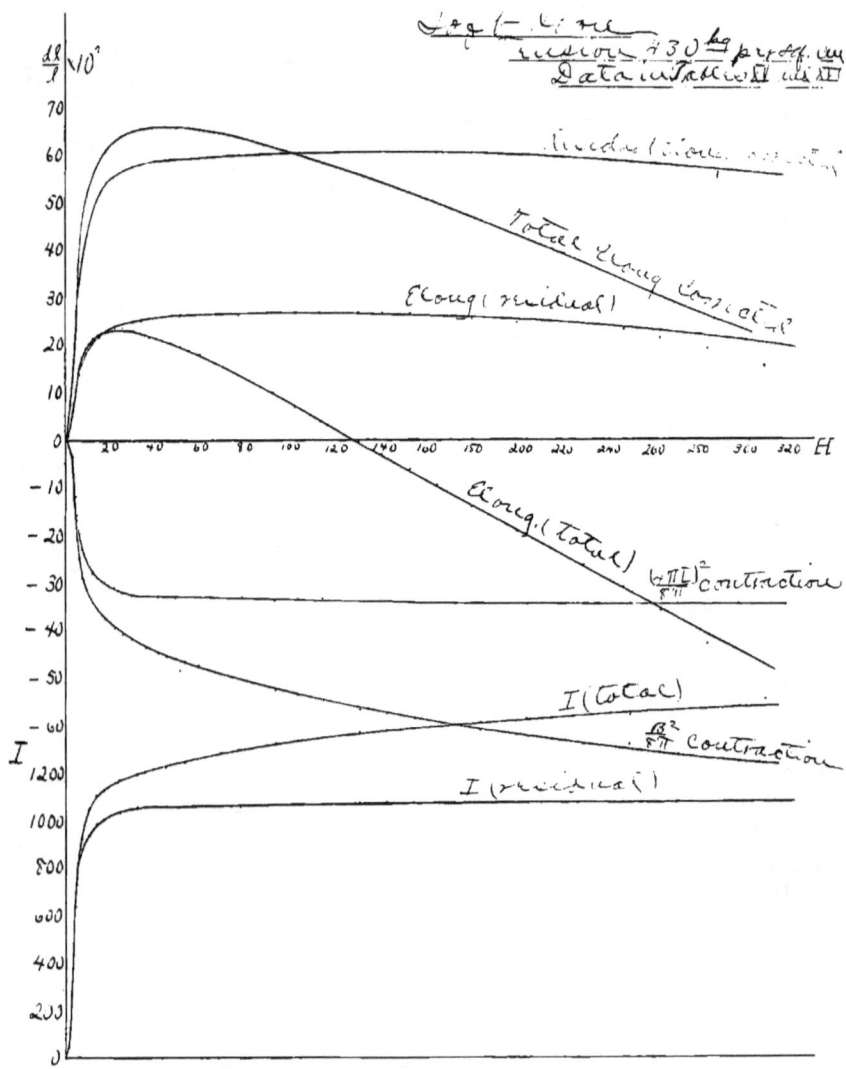

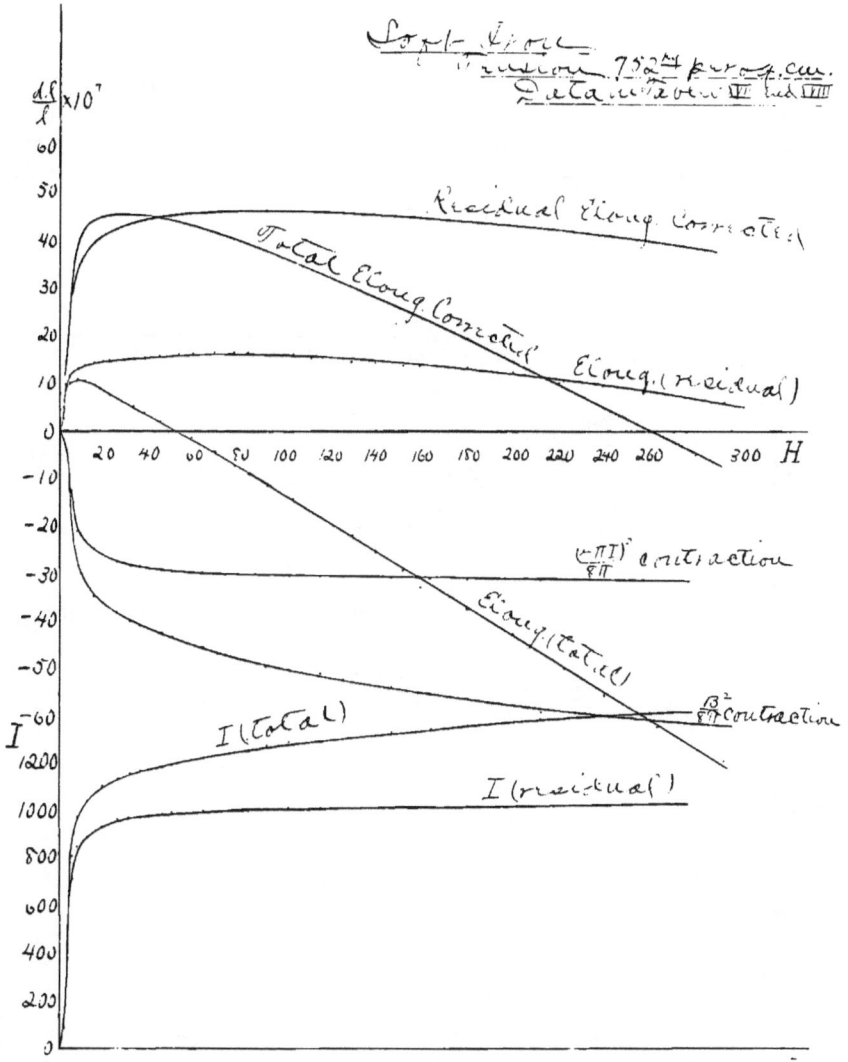

Soft Iron
Tension 752ᵈ per sq. cm.
Data as between ⅦI and Ⅷ

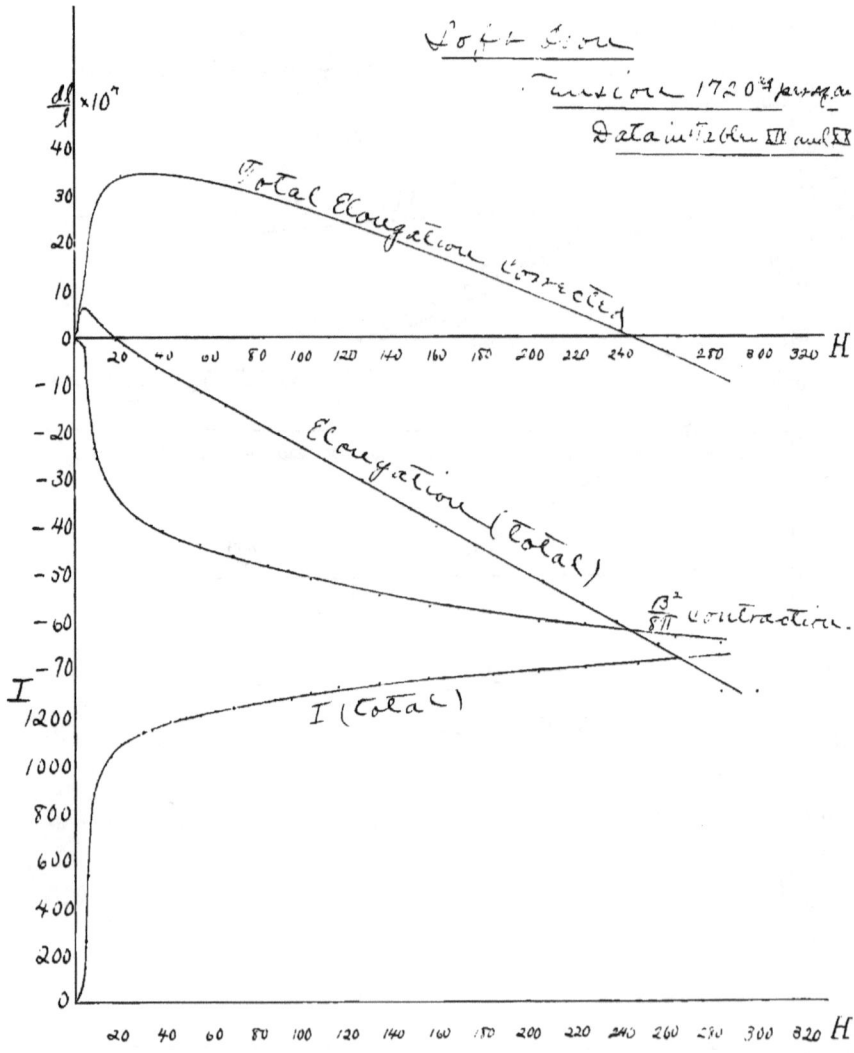

Soft Iron

Tension 1720 gr. per sq. cm

Data in Tables VI and VII

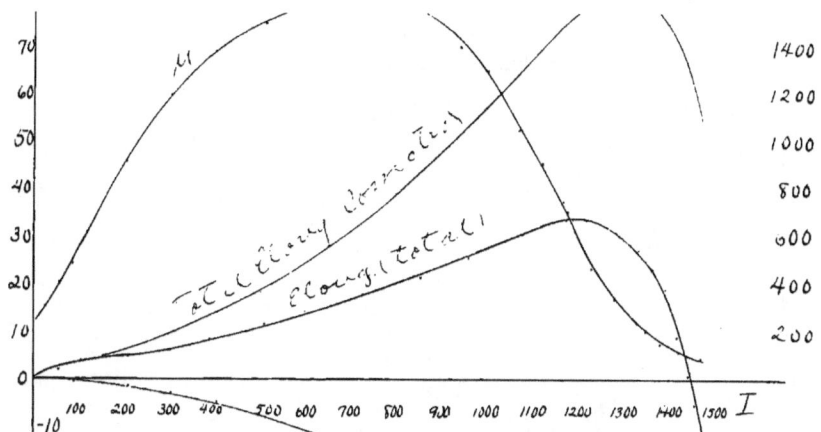

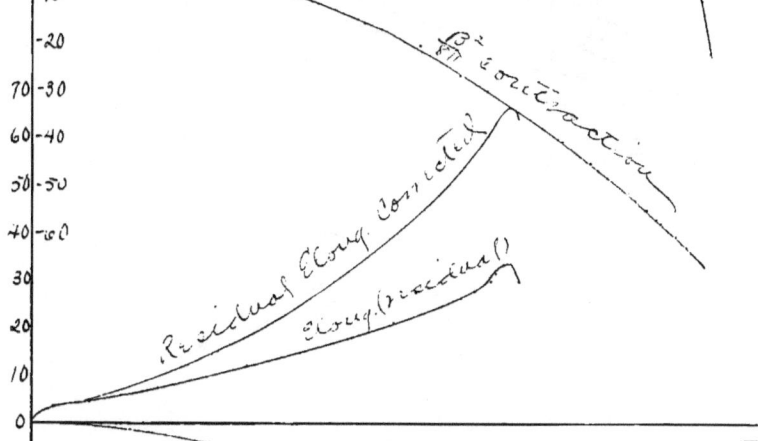

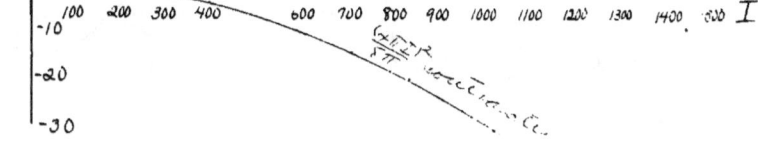

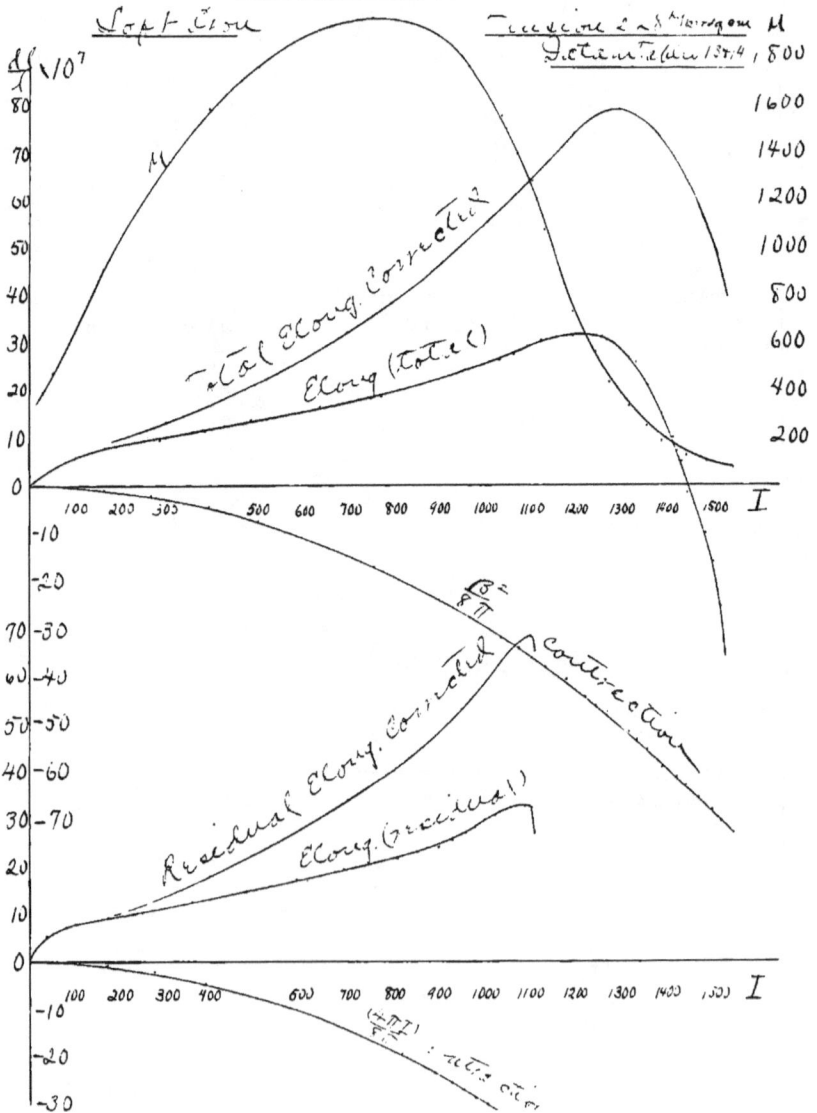

Plate VII

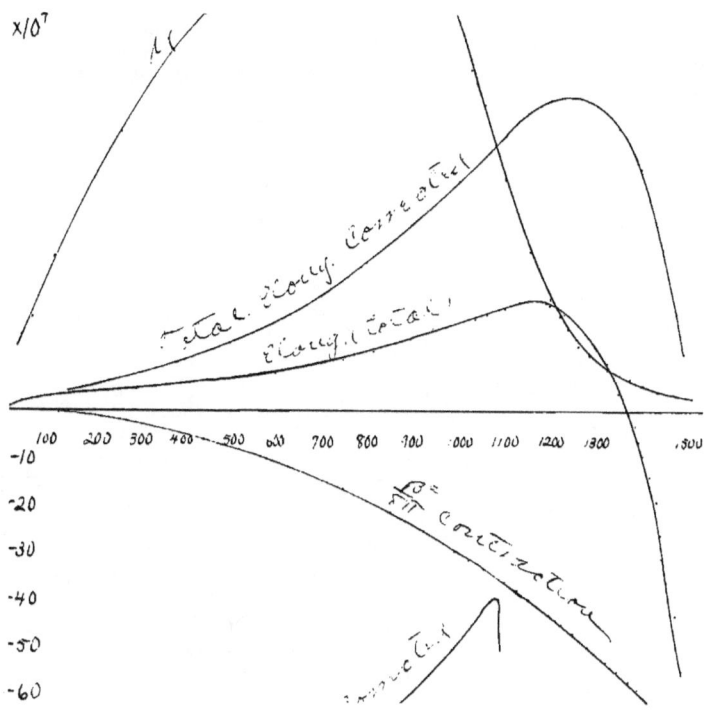

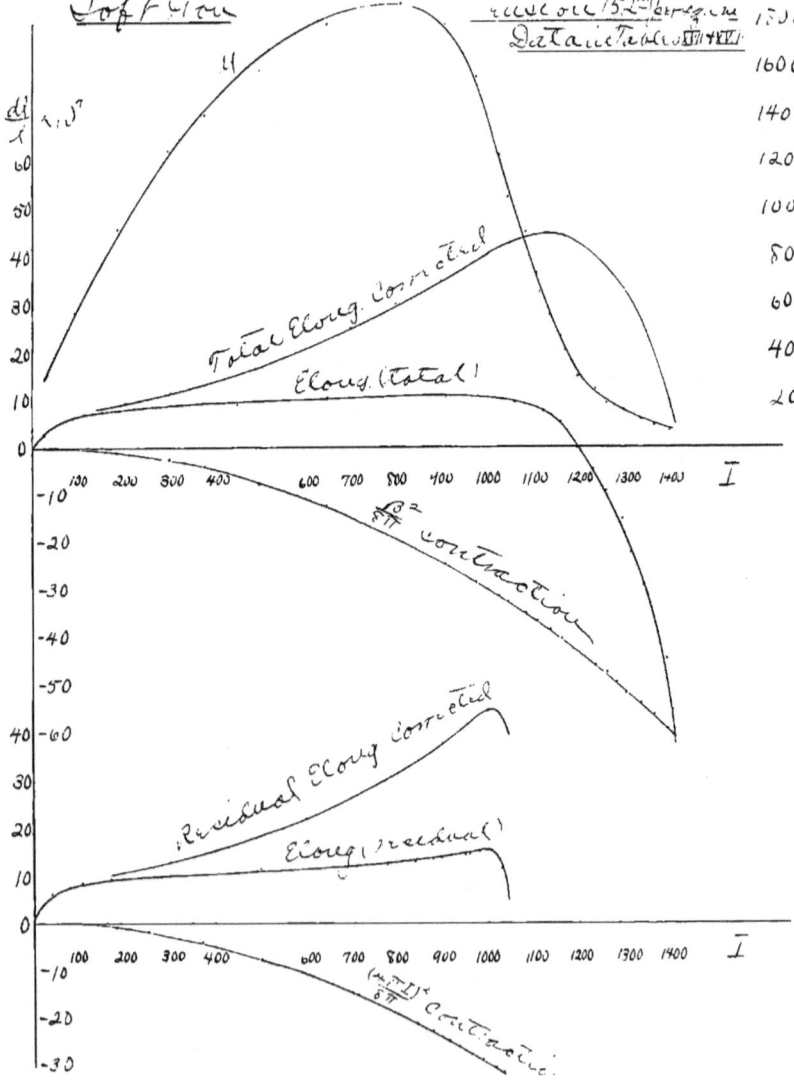

Soft Iron

Trace on 75.2 kilograms.
Data in Table XII + XIII

$\frac{di}{1} \times 10^7$

Total Elong. corrected

Elong. (total)

$\frac{\pi \theta^2}{8\pi}$ contraction

Residual Elong corrected

Elong. (residual)

$\left(\frac{\pi \pi - 1}{8\pi}\right)^2$ Contraction

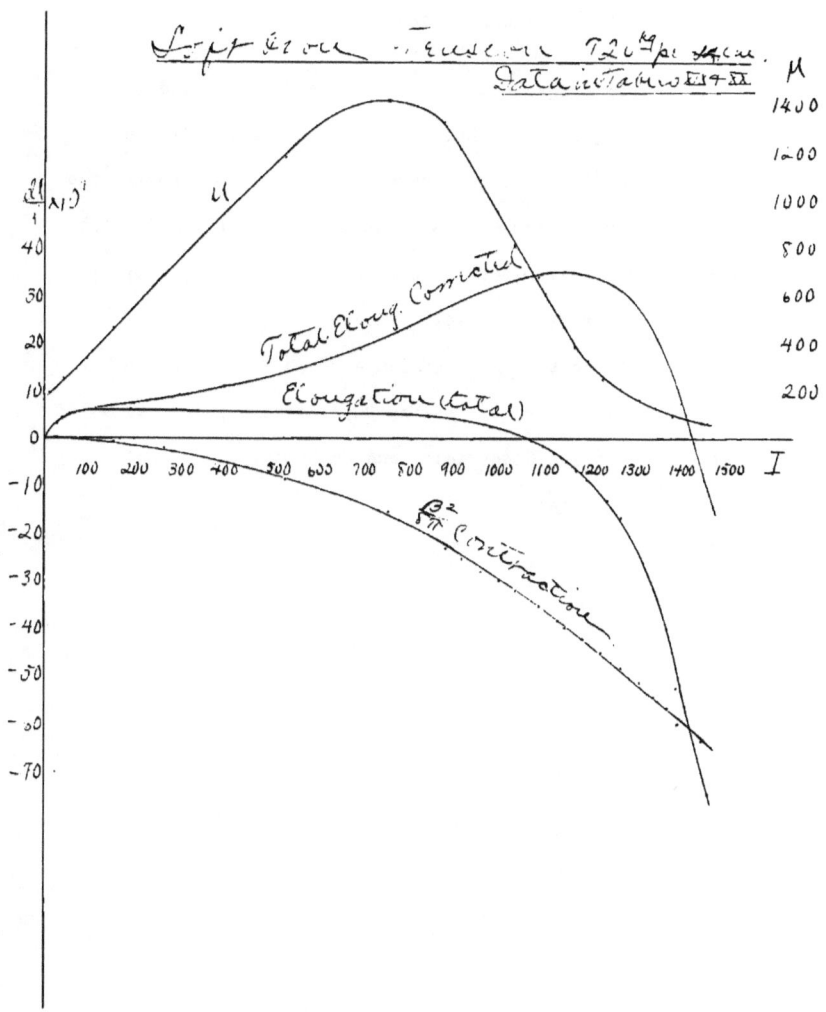

BIOGRAPHICAL.

Byron Briggs Brackett was born August 13, 1865, in
Cayuga County, New York State. He completed a collegiate
course at the Syracuse University in June, 1890. During the
following three years he was engaged in teaching, being two
years at the Williamsport Dickinson Seminary, Williamsport,
Pa., and one year at the Adelphi Academy and College of
Brooklyn. Since then he has been a graduate student in the
Johns Hopkins University, taking Physics as a principal subject, with Applied Electricity and Mathematics as first and
second subordinates.

www.ingramcontent.com/pod-product-compliance
Lightning Source LLC
Chambersburg PA
CBHW022005190326
41519CB00010B/1391

* 9 7 8 3 3 3 7 1 2 5 3 6 3 *